ENERGY SECTOR STANDARD OF THE PEOPLE' S REPUBLIC OF CHINA

中华人民共和国能源行业标准

Specification for Safety Appraisal of Hydropower Engineering

水电工程安全鉴定规程

NB/T 35064-2015

Chief Development Department: China Renewable Energy Engineering Institute

Approval Department: National Energy Administration of the People's Republic of China

Implementation Date: March 1, 2016

China Water & Power Press

中国水利水电出版社

Beijing 2024

图书在版编目（CIP）数据

水电工程安全鉴定规程 : NB/T 35064-2015 = Specification for Safety Appraisal of Hydropower Engineering（NB/T 35064-2015） : 英文 / 国家能源局发布. -- 北京 : 中国水利水电出版社, 2024. 10.
ISBN 978-7-5226-2796-0

Ⅰ. TV513-65

中国国家版本馆CIP数据核字第20247PH884号

ENERGY SECTOR STANDARD
OF THE PEOPLE'S REPUBLIC OF CHINA
中华人民共和国能源行业标准

Specification for Safety Appraisal of Hydropower Engineering

水电工程安全鉴定规程

NB/T 35064-2015

（英文版）

Issued by National Energy Administration of the People's Republic of China
国家能源局　发布
Translation organized by China Renewable Energy Engineering Institute
水电水利规划设计总院　组织翻译
Published by China Water & Power Press
中国水利水电出版社　出版发行

Tel: (+ 86 10) 68545888　68545874
sales@mwr.gov.cn
Account name: China Water & Power Press
Address: No.1, Yuyuantan Nanlu, Haidian District, Beijing 100038, China
http: //www.waterpub.com.cn

中国水利水电出版社微机排版中心　排版
北京中献拓方科技发展有限公司　印刷
184mm×260mm　16 开本　3.5 印张　111 千字
2024 年 10 月第 1 版　2024 年 10 月第 1 次印刷
Price（定价）: ¥ 450.00

Introduction

This English version is one of China's energy sector standard series in English. Its translation was organized by China Renewable Energy Engineering Institute authorized by National Energy Administration of the People's Republic of China in compliance with relevant procedures and stipulations. This English version was issued by National Energy Administration of the People's Republic of China in Announcement [2023] No. 8 dated December 28, 2023.

This version was translated from the Chinese Standard NB/T 35064-2015, *Specification for Safety Appraisal of Hydropower Engineering*, published by China Electric Power Press. The copyright is reserved by National Energy Administration of the People's Republic of China. In the event of any discrepancy in the implementation, the Chinese version shall prevail.

Many thanks go to the staff from the relevant standard development organizations and those who have provided generous assistance in the translation and review process.

For further improvement of the English version, any comments and suggestions are welcome and should be addressed to:

China Renewable Energy Engineering Institute
No. 2 Beixiaojie, Liupukang, Xicheng District, Beijing 100120, China
Website: www.creei.cn

Translating organization:

China Renewable Energy Engineering Institute

Translating staff:

YU Jialin	LI Xiang	CHENG Zhengfei	CHENG Li
KANG Wenjun	HE Jialong	LIU Biao	HOU Shaokang
YUE Lei			

Review panel members:

GUO Jie	POWERCHINA Beijing Engineering Corporation Limited
QIE Chunsheng	Senior English Translator
YAN Wenjun	Army Academy of Armored Forces, PLA
LI Zhongjie	POWERCHINA Northwest Engineering Corporation Limited

CHEN Lei	POWERCHINA Zhongnan Engineering Corporation Limited
YE Bin	POWERCHINA Huadong Engineering Corporation Limited
JIANG Jinzhang	POWERCHINA Huadong Engineering Corporation Limited
QI Wen	POWERCHINA Beijing Engineering Corporation Limited
WEI Fang	China Renewable Energy Engineering Institute
CHE Zhenying	IBF Technologies Co., Ltd.

National Energy Administration of the People's Republic of China

翻译出版说明

本译本为国家能源局委托水电水利规划设计总院按照有关程序和规定，统一组织翻译的能源行业标准英文版系列译本之一。2023 年 12 月 28 日，国家能源局以 2023 年第 8 号公告予以公布。

本译本是根据中国电力出版社出版的《水电工程安全鉴定规程》NB/T 35064—2015 翻译的，著作权归国家能源局所有。在使用过程中，如出现异议，以中文版为准。

本译本在翻译和审核过程中，本标准编制单位及编制组有关成员给予了积极协助。

为不断提高本译本的质量，欢迎使用者提出意见和建议，并反馈给水电水利规划设计总院。

地址：北京市西城区六铺炕北小街 2 号
邮编：100120
网址：www.creei.cn

本译本翻译单位：水电水利规划设计总院

本译本翻译人员：喻葭临 李 祥 程正飞 程 立 康文军 何佳龙 刘 彪 侯少康 岳 蕾

本译本审核人员：

郭 洁 中国电建集团北京勘测设计研究院有限公司

郄春生 英语高级翻译

闫文军 中国人民解放军陆军装甲兵学院

李仲杰 中国电建集团西北勘测设计研究院有限公司

陈 蕾 中国电建集团中南勘测设计研究院有限公司

叶 彬 中国电建集团华东勘测设计研究院有限公司

江金章 中国电建集团华东勘测设计研究院有限公司

齐 文 中国电建集团北京勘测设计研究院有限公司

魏 芳 水电水利规划设计总院

车振英 一百分信息技术有限公司

国家能源局

Announcement of National Energy Administration of the People's Republic of China [2015] No. 6

According to the requirements of Document GNJKJ [2009] No. 52, "Notice on Releasing the Energy Sector Standardization Administration Regulations (*tentative*) and detailed implementation rules issued by National Energy Administration of the People's Republic of China", 96 energy sector standards (NB) such as *Specification of Shale Gas Reservoir Description* are issued by National Energy Administration of the People's Republic of China after due review and approval.

Attachment: Directory of Sector Standards

National Energy Administration of the People's Republic of China

October 27, 2015

Attachment:

Directory of Sector Standards

Serial number	Standard No.	Title	Replaced standard No.	Adopted international standard No.	Approval date	Implementation date
…						
37	NB/T 35064-2015	Specification for Safety Appraisal of Hydropower Engineering			2015-10-27	2016-03-01
…						

Foreword

According to the requirements of Document GNKJ [2010] No. 320 issued by National Energy Administration of the People's Republic of China, "Notice on Releasing the Development and Revision Plan of the First Batch of Energy Sector Standards in 2010", and after extensive investigation and research, summarization of practical experience in the safety appraisal of hydropower projects, and wide solicitation of opinions, the drafting group has prepared this specification.

The main technical contents of this specification include: purpose and function of safety appraisal, main basis, classification, organization, procedure, assessment content and depth requirements, and responsibilities of and requirements for project participants in the safety appraisal.

National Energy Administration of the People's Republic of China is in charge of the administration of this specification. China Renewable Energy Engineering Institute has proposed this specification and is responsible for its routine management. Energy Sector Standardization Technical Committee on Hydropower Investigation and Design is responsible for the explanation of specific technical contents. Comments and suggestions in the implementation of this specification should be addressed to:

China Renewable Energy Engineering Institute
No. 2 Beixiaojie, Liupukang, Xicheng District, Beijing 100120, China

Chief development organization:

China Renewable Energy Engineering Institute

Participating development organizations:

POWERCHINA Limited

HydroChina Corporation

Chief drafting staff:

CHEN Huiming	ZHENG Xingang	CHANG Zuowei	YU Qinggui
WANG Runling	WEI Xiaowan	ZHAO Quansheng	YANG Baiyin
DAI Kangjun	GONG Jianxin	GUO Decun	LIN Zhaohui
LI Xiushu	SU Liqun	LIU Guoyang	LI Fuyun
LI Guangshun	FANG Hui	WU Yijin	YU Jialin
HOU Hongying			

Review panel members:

WANG Minhao	ZHOU Jianping	WEI Zhiyuan	ZHAO Kun
PENG Caide	LI Shisheng	YANG Zhigang	WAN Wengong
YUAN Dingyuan	CHEN Xuede	YANG Duogen	XIE Xiaoyi
ZHENG Zixiang	WEN Yanfeng	AI Yongping	YAN Jun
LYU Mingzhi	WANG Renkun	ZHANG Zongliang	ZHOU Chuiyi
YUAN Lianjun	SHI Qingchun	CHEN Yinqi	CHEN Hong
WAN Zongli	WAN Tianming		

Contents

1 General Provisions

1.0.1 This specification is formulated with a view to standardizing the safety appraisal of hydropower engineering, to make the scientific assessment on the project safety and improve the appraisal quality.

1.0.2 This specification is applicable to the construction, extension and renovation hydropower projects.

1.0.3 The safety appraisal of hydropower projects shall be independent, objective and science-based.

1.0.4 In addition to this specification, the safety appraisal of hydropower projects shall comply with other current relevant standards of China.

2 General Requirements

2.1 Classification and Overall Requirements

2.1.1 The safety appraisal is classified into special safety appraisal, impoundment safety appraisal, and completion safety appraisal. The special safety appraisal shall be completed respectively before the river closure acceptance, turbine-generator unit start-up acceptance and special works acceptance. Other special safety appraisal shall be completed before the corresponding works are put into operation. The impoundment safety appraisal and completion safety appraisal shall be completed before the impoundment acceptance and hydropower complex acceptance, respectively.

2.1.2 The basis documents for safety appraisal shall include the following:

1 Relevant state laws and regulations, and sector regulations.

2 National and sector specifications, codes and technical standards.

3 Project examination, approval, and registration documents.

4 Reviewed or approved project feasibility study report, special report, major design change report and their review comments.

5 Contract documents and construction drawing design documents related to the project construction.

2.1.3 The safety appraisal shall identify the problems affecting the project safety in design, construction, equipment testing, and initial operation as per the relevant basis documents, and propose the project safety assessment comments.

2.1.4 The safety appraisal report shall give a clear and definite conclusion. The special safety appraisal shall draw a conclusion of whether the project has met the conditions for operation within the specified scope of safety appraisal; the impoundment safety appraisal shall draw a conclusion of whether the hydropower complex has met the conditions for impounding to the target level within the specified scope of safety appraisal; the completion safety appraisal shall draw a conclusion of whether the hydropower complex has met the conditions for normal operation.

2.1.5 The safety appraisal shall not involve any quality rating or appraising.

2.2 Special Safety Appraisal

2.2.1 The special safety appraisal shall be carried out according to the project needs in any of the following cases:

1 Before diversion release structures under complex conditions are put into operation for water passing.

2 For the project using the staged diversion, the permanent structures and concealed works constructed at the early stage need to be put into operation before the impoundment safety appraisal.

3 Before the important water conveyance tunnels are put into operation for water filling.

4 Where the locations are found with safety hazards in the construction and initial operation.

5 When navigation structure, intake structure and other special works require separate acceptance.

2.2.2 The special safety appraisal shall include the following:

1 Check and assess the design, construction quality, and safety monitoring conditions within the specified scope of safety appraisal, focusing on key locations for the construction quality and initial operation and other locations where the quality defects and events occurred, which, if necessary, shall be confirmed by technical means of detection and testing.

2 Assess the conditions under which the project can be put into safe operation within the scope of safety appraisal, and put forward the comments and suggestions on problems of concern to the future operation.

2.3 Impoundment Safety Appraisal

2.3.1 The impoundment safety appraisal shall be centered on the dam, focusing on the impoundment safety-related items, including the civil works, safety monitoring works and relevant hydraulic steel structure works of water retaining structure, water and sediment release structures, inlet and outlet of water conveyance structure, seepage control works, downstream energy dissipation protection structures, reservoir bank near dam, etc.

2.3.2 For the impoundment safety appraisal, the following conditions shall be available:

1 The physical progress of the dam and other water retaining structures, and their foundation treatment and seepage control works, has met the impoundment requirements, and the subsequent construction and flood control will not be affected after the impoundment.

2 The inlet and outlet of the water conveyance structure have been

basically completed, the gates and hoists for retaining water have been installed and tested, and the power supply is reliable.

3 The water release structure to be put into operation after the impoundment has been basically completed, with water discharge conditions available, the gates and hoists for controlling the discharge have been installed and tested, and the power supply is reliable.

4 The monitoring instruments and equipment for the structures and engineering slopes related to impoundment have been embedded and the initial values have been measured.

5 The plugging gates, gate slots and hoists of diversion structures have been installed and tested, and the power supply is reliable.

6 The reservoir leakage, bank stability, immersion, debris flow, etc. affecting the safe operation of the hydropower complex after the impoundment have been handled according to the design requirements, and the monitoring facilities for reservoir-induced earthquake have been installed according to the design requirements.

2.3.3 The impoundment safety appraisal shall include the following:

1 Check and assess the project physical progress.

2 Check and assess the design, construction quality, and safety monitoring conditions within the specified scope of safety appraisal, focusing on key locations for the construction quality and other locations where the quality defects and events occurred, which, if necessary, shall be confirmed by technical means of detection and testing.

3 Assess the impoundment program of the project.

4 Assess the flood control scheme, flood control measures and emergency response plan of the project.

5 Assess the impoundment conditions of the hydropower complex, and put forward the comments and suggestions on problems of concern to impoundment.

2.3.4 The impoundment safety appraisal of the project with staged impoundment may be carried out in stages.

2.3.5 For the staged impoundment safety appraisal, the safety appraisal report of the later stage shall include the assessment on treatment of pending issues in the safety appraisal report of the previous stage, and on the conditions of the hydropower complex which has been put into operation since the impoundment.

2.3.6 The impoundment safety appraisal report shall serve as one of the basis for project impoundment acceptance.

2.4 Completion Safety Appraisal

2.4.1 The completion safety appraisal shall include the civil works, safety monitoring works, hydraulic steel structure, electromechanical works, etc. of water retaining structure, water release structure, power generation structure, seepage control works, dam abutments on both banks, reservoir bank slope near dam, etc. Navigation structure, intake structure, and other special works which have been put into operation before the completion safety appraisal shall be included in the specified scope of the completion safety appraisal.

2.4.2 For the completion safety appraisal, the following conditions shall be available:

1 The hydropower complex has been fully completed according to the approved design scale and criteria.

2 The relevant review procedures for major design changes have been completed.

3 Winding-up works and defects remedying have been completed.

4 The project operation has undergone at least one flood period, the multiyear regulation reservoir has undergone at least two flood periods; the highest reservoir water level has reached or basically reached the normal pool level; all units have operated at rated output; each unit in the conventional hydropower station has operated for at least 2000 h, and each unit in the pumped storage power station has operated for at least 800 h.

2.4.3 The completion safety appraisal shall include the following:

1 Check and assess the project physical progress.

2 Check and assess the design and construction quality of the hydropower complex, focusing on key locations for the construction quality and other locations where the quality defects and events occurred, which, if necessary, shall be confirmed by technical means of detection and testing.

3 Analyze and assess the initial operation condition of the hydropower complex.

4 Check and assess the problems identified with the design, construction and initial operation of the hydropower complex and the treatment results.

5 Comprehensively assess the safety and reliability of the hydropower complex under normal operation, and put forward the comments and suggestions on problems of concern to operation.

2.4.4 For the completion safety appraisal, the hydropower complex may be divided into several parts as needed, and their appraisals shall be combined into a completion safety appraisal report before the special acceptance of the hydropower complex. If necessary, the safety appraisal for civil works and electromechanical works of the hydropower complex may be carried out independently. The completion safety appraisal report shall review the main specialized comments given in the special safety appraisal and impoundment safety appraisal, and provide the supplementary assessment on the operation conditions of the corresponding main structures under operation.

2.4.5 For the project construction in stages, the safety appraisal may be conducted in stages or once according to the actual situation of the project construction. If done in stages, safety appraisal reports may be independent of each other.

2.4.6 The completion safety appraisal report shall serve as one of the basis for special acceptance of the hydropower complex and the project safety management after it enters into normal operation.

3 Organization and Procedure

3.1 Organization

3.1.1 The safety appraisal shall be undertaken by an agency that is entrusted by the project owner and possesses relevant experience in the safety appraisal of hydropower projects.

3.1.2 The safety appraisal agency shall be responsible for establishing an expert team to carry out the safety appraisal.

3.1.3 The expert team shall consist of senior experts in planning, geology, hydraulic engineering, construction, safety monitoring, hydraulic steel structure, electromechanical engineering, and other related disciplines. The personnel who have been involved in the construction management, design, construction, supervision, safety monitoring, equipment manufacture, installation, operation, etc. of the project shall not serve as a member of the expert team.

3.1.4 The safety appraisal agency shall be cooperative when the competent department of hydropower project safety appraisal oversees the safety appraisal process. The competent department assesses the quality of the final safety appraisal report and examines the performance of the safety appraisal agency.

3.2 Procedure

3.2.1 Safety appraisal shall include the following four stages:

1 Prepare the safety appraisal program.

2 Conduct the site appraisal.

3 Prepare the safety appraisal report.

4 Report to the acceptance committee on the safety appraisal and the main conclusions.

3.2.2 The preparation of the safety appraisal program shall include the following:

1 The safety appraisal expert team shall conduct the on-site investigation, and listen to the reports from parties involving the project development, design, construction, supervision, third-party inspection, safety monitoring, equipment manufacture and installation, operation, etc.

2 Determine the work focus and requirements, and define the appraisal scope and main content.

3 Specify the documents to be prepared or provided by project participants for safety appraisal; put forward the required additional calculation review, and inspection and testing tasks by collecting and collating the possible problems affecting the construction safety in the aspects such as design, construction, and operation; define the requirements for the preparation of self-inspection reports of project participants.

4 Prepare and issue the safety appraisal program. The contents of the safety appraisal program may be supplemented and improved in accordance with the relevant contents listed in Appendix A of this specification.

3.2.3 The expert team shall carry out the on-site appraisal based on the safety appraisal program and project progress, and comprehensively understand the project construction, sort out and analyze main problems through on-site inspection, information consulting, listen to briefings and talk with the project participants; check their self-inspection reports according to the relevant requirements of this specification, and put forward the improvement suggestions; require the relevant participants to supplement necessary demonstration, test verification or inspection and testing results in respect of the problems identified with the design, construction quality, initial operation, etc.; assess the project quality according to the requirements specified in basis documents mentioned in Article 2.1.2 of this specification if the conditions are right.

3.2.4 The expert team shall draft the safety appraisal report through the analysis of project information and study of key issues; the safety appraisal report may be prepared in accordance with Appendix B of this specification.

3.2.5 The expert team shall draw the main conclusions of the safety appraisal report based on the exchange of opinions with the project participants.

3.2.6 The safety appraisal report shall be signed by all members of the expert team, approved by the legal representative of the safety appraisal agency, and stamped and then submitted to the project owner. If approved by a person other than the legal representative of the safety appraisal agency, the report shall be accompanied by the corresponding power of attorney signed by the legal representative.

3.2.7 Safety appraisal agency shall report the safety appraisal work and the main conclusions based on the acceptance arrangements to the acceptance committee. According to the requirements of the acceptance presiding organization or acceptance committee, the safety appraisal report shall be supplemented and revised, if necessary, and resubmitted to the project owner.

3.2.8 After the completion of the safety appraisal, the safety appraisal report and the main basis documents shall be archived in accordance with relevant requirements.

4 Responsibilities and Requirements for All Parties

4.0.1 The project owner shall, according to the relevant provisions on the safety appraisal, acceptance management, etc. of hydropower projects, develop a safety appraisal schedule, and incorporate it into the project development plan.

4.0.2 The project owner shall organize and coordinate all parties involving the project development, design, supervision, construction, third-party inspection, safety monitoring, equipment manufacture and installation, and operation, to assist the safety appraisal agency and provide the information required. The documents and information required shall be in accordance with Appendix C of this specification.

4.0.3 All participants shall, according to the safety appraisal program and other requirements of the safety appraisal expert team, prepare and supplement the corresponding safety self-inspection reports from parties involving project development, design, supervision, construction, third-party inspection, safety monitoring, equipment manufacture and installation, operation within the specified scope of safety appraisal. The format of the self-inspection report shall be in accordance with Appendix D of this specification. The self-inspection report shall be examined by the technical chief of the organization, and stamped with the official seal of the organization.

4.0.4 For the problems raised by the safety appraisal expert team, the participants shall conduct additional analysis, and provide the relevant additional information, and if necessary, submit the special reports; the project owner shall organize the participants to study the main problems and comments in the safety appraisal report, and put forward the rectification plan and measures and implement them.

4.0.5 The participants shall be responsible for the integrity of all information they provide. Any organization that does not report or gives a false report on the safety hazards in the project shall be held accountable for such acts in accordance with the relevant provisions.

4.0.6 The participants shall not hinder and interfere with the safety appraisal agency and safety appraisal expert team from making their independent opinions.

4.0.7 In the case of any significant disagreement with the safety appraisal report, the participants shall submit a written opinion letter to the safety appraisal agency, and copy to the acceptance presiding organization and the acceptance committee.

4.0.8 The project owner shall submit the safety appraisal report to the acceptance presiding organization, and copy to the members of the acceptance committee and the acceptance expert team at the time of the project acceptance.

4.0.9 If the acceptance presiding organization or the acceptance committee thinks that the safety appraisal report does not meet the relevant requirements of this specification or acceptance requirements, the project owner shall assist the safety appraisal agency in revising the report as required.

4.0.10 The project safety appraisal work does not substitute and relieve the quality and safety responsibilities of the parties involving project development management, design, supervision, construction, third-party inspection, safety monitoring, equipment manufacture and installation, and operation.

5 Assessment Contents of Safety Appraisal

5.1 General Assessment

5.1.1 The conformity of the project construction to the national capital construction procedure shall be reviewed and assessed; the construction scale and function of the project shall be reviewed and assessed according to the project examination or approval documents.

5.1.2 The conformity of project rank, structure grade, flood control design criteria, seismic design criteria, main dimensions and key design parameters of structures, type selection, layout and key parameters of main equipment, etc. shall be assessed according to the reviewed feasibility study report, special design report, major design change report and the corresponding design contents approved or examined.

5.1.3 The project current physical progress shall be checked and assessed according to the design requirements.

5.1.4 The quality management organizational structure, management regulations, management procedures, etc. shall be checked to assess the project quality management system.

5.1.5 The project design shall be assessed against the current standards applicable to the safety appraisal stage. The design review results obtained according to the current design standards shall be assessed when the design standard adopted by the completed project has been repealed.

5.1.6 The quality of project construction and main equipment workmanship and installation shall be assessed according to the requirements of current standard, design or contract documents, and the results of safety review or testing verification proposed according to the actual specification parameters shall be checked as required.

5.1.7 For items within the specified scope of safety appraisal, remedy of the problems identified in the previous appraisal or stage acceptance shall be checked and assessed if the pending issues and treatment suggestions have been proposed in the previous safety appraisal within the specified scope or in the stage acceptance of the project.

5.2 Assessment on Impoundment, Flood Control and Operation

5.2.1 The extended hydrological data series shall be assessed for the representativeness, consistency, and reliability, and be compared with the runoff and design flood results used in the feasibility study stage. The reliability and rationality of the runoff and design flood calculation methods and review results after the hydrological series is extended shall be assessed. When necessary, the sediment results shall be assessed.

5.2.2 The capacity of flood discharge facilities shall be assessed according to the comparative analysis of capacity calculation and model test results of the flood discharge facilities.

5.2.3 The review results of the reservoir capacity curve shall be assessed according to the results of background measurement before the impoundment; the review results of downstream water stage-discharge relation shall be assessed.

5.2.4 The actual flood control capacity of the project shall be assessed through the comparative analysis of the design flood control characteristic water level based on review results of the design flood, reservoir capacity curve and discharge curve, and on the principles of flood regulation.

5.2.5 The establishment, operation parameters, and operation performance of hydrological telemetry and forecasting system shall be assessed according to the design requirements for smoothness, availability, forecasting programs, forecasting accuracy, etc.

5.2.6 In addition to Articles 5.2.1 to 5.2.5 of this specification, the following shall be conducted for the impoundment safety appraisal:

1 Assess the rationality of initial impoundment time, gate-closing procedures of diversion facilities, staged impoundment principles, water level control of staged impoundment, physical progress before the impoundment and its requirements, and design dependability of impoundment, and assess the initial impoundment plan of the project.

2 Assess the coverage of investigation of the downstream water supply scope affected by impoundment and the rationality of the influence analysis according to the analysis results of impoundment influence on the downstream water supply, ecological environment, power generation, shipping, etc., and assess the rationality of downstream water supply measures during impoundment.

3 Assess the rationality of requirements for impoundment procedure, control level and rise rate of main structures during impoundment according to the construction status of each structure and the stability requirements of landslide, bank collapse, etc. within the reservoir area.

4 Assess the feasibility and rationality of the impoundment plan according to the principles, control level, impoundment period, reservoir level rise rate, downstream water use proposed in the initial impoundment plan. For the initial impoundment plan of pumped storage power stations, the rationality and reliability of the reservoir replenishment plan shall be assessed according to the water source conditions.

5 Assess the conformity of the flood control criteria adopted by each flood control object in different periods to relevant specifications, according to their grades and flood control criteria during impoundment.

6 Assess the conformity of flood discharge facilities to the flood control requirements, according to their physical progress, discharge capacity and restrictions, as well as the gate operation rules during the impoundment.

7 Assess the starting level of flood regulation during impoundment in different periods, selected design flood hydrograph, flood regulation rules in different periods, as well as the rationality and reliability of the flood regulation results, and if necessary, assess the review results of the reservoir backwater surface profile as the basis for the staged resettlement.

8 Assess the feasibility of the flood control scheme, measures and exceeding-standard flood emergency response plan of the project during the initial impoundment according to the flood control standard, flood regulation results, physical progress and construction schedule, taking into account the implementation of the flood control scheme and flood control measures.

5.2.7 In addition to Articles 5.2.1 to 5.2.5 of this specification, the following shall be conducted for the completion safety appraisal:

1 Assess the effectiveness of reservoir regulation scheme and improvement measures after the initial impoundment with respect to the problems identified with the initial operation of the reservoir and their solutions.

2 Assess the capacity of the flood discharge facilities according to the gate operation procedures of the permanent flood discharge facilities and the treatment effect of the existent problems, taking into account the difference between the prototype observation results and the design results of the capacity of the flood discharge facilities in the operation stage.

3 Assess the operational safety and reliability of each of the multiple purposes of the project according to the regulation and operation rules of the hydropower station and reservoir, taking into account the operation effect of power generation, flood control (ice prevention), shipping, water supply, sediment retaining and releasing, etc.

4 Assess the rationality and feasibility of the flood emergency response plan and measures for the hydropower reservoir.

5.3 Assessment on Engineering Geological Conditions

5.3.1 The assessment on engineering geological conditions shall include the regional tectonic stability and the engineering geological conditions of the reservoir area and project area.

5.3.2 The assessment on regional tectonic stability shall meet the following requirements:

1 Check the active fault distribution and seismic hazards in the project area and the main research results, conclusions or relevant approval of ground motion parameters, seismic basic intensity, etc. in the dam area, and assess the regional tectonic stability of the project.

2 For the project requiring seismic safety review and assessment, propose comments and suggestions on implementation.

5.3.3 The assessment on engineering geological conditions of the reservoir area shall meet the following requirements:

1 Assess the stability review and analysis results of the reservoir bank near dam emphatically before the impoundment, including the stability of landslides, soil deposit, and potential instable bank slope that might affect the project safety, etc., and assess the stability review and analysis results of permanent reservoir bank in the completion safety appraisal according to the changes in the stability of reservoir bank slope after the impoundment.

2 For the project with reservoir leakage problems, assess the rationality and reliability of treatment scope of and measures for reservoir leakage according to the predictive analysis results of reservoir leakage conditions before the impoundment, and assess the review results of reservoir leakage in the completion safety appraisal according to the monitoring status of reservoir leakage after the impoundment.

3 Assess the possibility of reservoir-induced earthquake and the predictive analysis results of its impact on the project before the

impoundment. For reservoirs requiring seismic monitoring network, assess the design and implementation of the reservoir-induced earthquake monitoring network for reservoirs before the impoundment; assess the review and analysis results of change rules of earthquake before and after the impoundment, and the occurrence of reservoir-induced earthquake and its impact on the project in the completion safety appraisal according to the seismic monitoring data in the reservoir area after the impoundment.

5.3.4 The assessment on engineering geological conditions in the hydropower complex area shall meet the following requirements:

1 Assess the engineering geological classification of rock mass, recommended values of physical and mechanical parameters of rock and soil mass, and the adjustment rationality according to the basic geological conditions revealed and the comparative analysis of the early-stage investigation results.

2 Assess the engineering geological conditions of the foundation of the water retaining structures before the impoundment, including the selection rationality of the foundation surface, the stability against sliding, deformation and seepage of the dam foundation or abutment, the geological defects treatment measures and effects, etc.; assess the review results of foundation stability in the completion safety appraisal according to the operation and monitoring analysis results of the ground of all water retaining structures.

3 Assess the engineering geological conditions of the water release structures before the impoundment, including the stability of the water release structure foundation, cavern surrounding rocks, energy dissipation structure foundation and slopes of downstream energy dissipation and scour-resistant area and atomization area, the geological defects treatment measures and effects, etc.; assess the review results of engineering geological conditions, stability, etc. of the water release structure foundation, cavern surrounding rocks, energy dissipation structure foundation and slopes of downstream energy dissipation and scour-resistant area and atomization area in the completion safety appraisal after the impoundment.

4 Assess the engineering geological conditions of the ground of the surface powerhouse and other structures before the impoundment as needed. Assess the review results of stability of ground or surrounding rocks and the seepage stability, and the geological conditions and stability review results of the switchyard ground and slopes in the completion safety appraisal according to the engineering geological conditions and monitoring results of surface powerhouse or underground powerhouse caverns and water conveyance system caverns.

5 Assess the slope stability affected by the impoundment according to the basic geological conditions, deformation mechanism and failure mode, initial monitoring results, etc. before the impoundment; assess the effectiveness of slope support and the slope stability in the completion safety appraisal based on the monitoring results and stability review results of slopes.

6 Assess the rationality and reliability of seepage treatment scope according to the revelation by drilling for curtain grouting and the hydrogeological and engineering geological conditions of the dam area before the impoundment; assess the seepage treatment effect of dam foundation in the completion safety appraisal according to the changes in hydrogeological conditions, monitoring of seepage flow and seepage pressure, etc. after impoundment.

7 Assess the engineering geological conditions of surrounding rocks in the plugging sections of the diversion tunnel, main access tunnel connected with the reservoir, main construction adits and other channels before the impoundment; assess the stability review results of surrounding rocks in the plugging sections in the completion safety appraisal according to the monitoring of seepage flow, seepage pressure, etc. after the impoundment.

8 Assess the stability of landslide, deposit, deformation body, and natural slope dangerous rock mass affected by the impoundment near the project area and their impact on the project safety and the control effect before the impoundment; assess the danger of debris flow to the project safety and its control effect; in the completion safety appraisal according to the monitoring data; assess the stability of landslide, deposit, deformation body, natural side slope dangerous rock mass and other potential unstable rock and soil mass near the project area, and their impact on the project safety and the control effect in the completion safety appraisal.

5.4 Assessment on Hydraulic Structure Design

5.4.1 The assessment on hydraulic structure design includes the assessment on the design of the water retaining structure, water release structure, water conveyance and power generation structure, navigation structure, fish pass structure, seepage control works, slope works, diversion tunnel plugging works and other structures. The structure design shall meet the following requirements:

1 Assess the rationality of the values of main design parameters, especially the design parameters used when the actual engineering geological conditions and geological parameters have changed significantly compared to those at the feasibility study stage, according to the review results of basic data on hydrology, meteorology, geology, etc.

2 Assess the treatment schemes and design treatment standards for quality defects and events occurring in the construction process; if necessary, assess the rationality of the corresponding design parameters used according to the actual treatment effect.

3 Assess the treatment schemes of the problems and defects identified in the initial operation of the structures and slopes and their treatment effectiveness, assess the operation safety of the structures and slopes comprehensively based on the analysis results of safety monitoring data and the initial operation, and put forward comments and suggestions on the problems to be addressed in the future operation in the completion safety appraisal.

5.4.2 The assessment on water retaining structure design shall meet the following requirements:

1 For concrete gravity dams, assess the design of dam foundation treatment, dam crest elevation, dam section, dam concrete materials and zoning, dam gallery layout, dam jointing structure and drainage, etc.; assess the sliding stability of dam foundation, roller compacted concrete layer and deep layer of dam foundation as well as dam foundation bearing capacity under various operating conditions; assess the dam concrete temperature control design.

2 For concrete arch dams, assess the design of dam foundation treatment, dam crest elevation, arch dam shape, dam concrete materials and zoning, dam gallery layout and other structures; assess the arch dam stress-induced deformation and abutment sliding stability under various operating conditions; assess the design of arch dam quasi-stable temperature field, arch closure temperature, concrete temperature control and joint grouting.

3 For dams built with local materials, assess the design of dam crest elevation, dam section, dam materials and zoning, and foundation treatment, and assess the dam slope sliding stability, and dam settlement deformation; and the dam seepage control design, and seepage analysis results.

4 For ultra-high dams exceeding the current design specifications and for dams on deep overburden, assess the major technical issues and special research results.

5 Assess the stability, safety, stress-induced deformation, etc. of the temporary water retaining section of the dam in thc impoundment safety appraisal.

5.4.3 The assessment on water release structure design shall meet the following requirements:

1 Assess the layout, shape, structure and foundation treatment design of the water release structure; assess the overall stability and safety of the gate control section, the internal force analysis and reinforcement design of the main structures such as concrete gate piers; assess the stability and support design of the underground cavern surrounding rocks, and the internal force analysis, reinforcement design, backfill grouting, contact grouting design, etc. of lining structures.

2 Assess the discharge capacity and the hydraulic characteristics of energy dissipation structures and assess the shapes of the flow channel and energy dissipator, taking into account the hydraulic design and hydraulic model test results.

3 Assess the layout, structural design and foundation treatment of energy dissipation structures, and assess the measures and structural design of downstream scour-resistant works.

4 Assess the operation mode of the water release structures and gates.

5.4.4 The assessment on water conveyance system design shall meet the following requirements:

1 Assess the layout, structural design, and foundation treatment of the intake and tailwater outlet, assess the internal force analysis and reinforcement design of the main structures, and assess the hydraulic design of the intake and outlet.

2 Assess the layout of the water conveyance system structures and the design, internal force analysis and reinforcement design of lining structures (including steel lining structure) according to the review results of the hydraulic transient process.

3 Assess the stability and support design of surrounding rocks in the water conveyance system caverns, and assess the design of consolidation grouting of surrounding rocks in the water conveyance tunnel, backfill grouting and contact grouting of the lining structure, etc.

5.4.5 The assessment on power generation structure design shall meet the following requirements:

1 Assess the layout of the powerhouse, and the layout, structural design and foundation treatment design of powerhouse machine hall, erection bay, auxiliary rooms, step-up substation, and other structures. For the surface powerhouse, assess the overall stability and foundation bearing capacity; for the underground powerhouse, assess the stability and support design of surrounding rocks in caverns.

2 Assess the structural design, internal force calculation, and rein forcement design of the powerhouse spiral case, rock-bolted crane girder, air housing, generator pier, draft elbow tube, and other major structures.

5.4.6 The assessment on navigation structure design shall meet the following requirements:

1 Assess the layout, structural design, and foundation treatment design of navigation structures; assess the hydraulic characteristics and operation conditions of navigation structures according to the research results of hydraulic model tests, model trials, etc., and assess the hydraulic characteristics and structural design of the water conveyance system.

2 Assess the internal force analysis and reinforcement design of the main reinforced concrete components of the navigation structure.

3 Assess the lock head of the navigation structure with water retaining requirements and the auxiliary building of the approach channel below the inundation line in the reservoir area in the impoundment safety appraisal, focusing on their overall stability, foundation bearing capacity, structural design, and foundation treatment design.

5.4.7 The assessment on fish pass structure design shall meet the following requirements:

1 Assess the layout and foundation treatment design of the fish pass structure, and assess its shape according to the hydraulic calculation and hydraulic structural model test results.

2 Assess the main structural design, internal force calculation, and reinforcement design of fish pass structure.

3 Assess the operation safety of the fish pass structure.

5.4.8 The assessment on seepage control works design shall meet the following requirements:

1 Assess the seepage control design criteria and seepage control scheme, and assess the seepage control works design based on thc construction and control effect of seepage control works.

2 Assess the design of the pumping and drainage system.

3 For pumped storage power stations, assess the seepage control criteria and scheme of the upper and lower reservoir basins.

5.4.9 The assessment on slope works design shall meet the following requirements:

1 Assess the stability, safety and the support measures of excavated slopes of permanent structures and the slopes affected by atomization, and assess the stability and support design of important temporary slopes.

2 Assess the stability, safety, and engineering treatment measures of the reservoir bank slopes with safety hazards near the dam.

3 Assess the stability and treatment design of natural slopes and dangerous rock mass around the project structures.

5.4.10 The plugging design of the diversion structures and the channels connected to the reservoir shall be assessed in the impoundment safety appraisal.

5.4.11 The assessment on construction quality defects or defect remedy design shall include:

1 Assess the cause analysis, treatment solutions and effects of quality defects and events and other issues occurring in the construction process, and put forward the comments and suggestions on the problems to be addressed in the impoundment process and in the impoundment safety appraisal.

2 Assess the cause analysis, treatment solutions and effects of the problems exposed in the initial operation, and put forward the comments and suggestions on the problems to be addressed in the future operation in the completion safety appraisal.

5.5 Assessment on Civil Works Construction Quality and Penstock Manufacture and Installation Quality

5.5.1 The operational effectiveness of the project quality control and management system shall be assessed in terms of the establishment of project quality management system and regulations, raw material supply, construction process quality control, inspection and acceptance, etc.

5.5.2 The physical progress of the civil works shall be assessed according to the physical progress design requirements, physical progress and quantities, acceptance certification status, uncompleted works schedule, guarantee measures, etc. of each section of works and part of works within the scope of safety appraisal assessment. The construction schedule rationality of the uncompleted works in the impoundment safety appraisal if the physical progress of important construction works such as water retaining and release structures and seepage control works might restrict the impoundment of the project shall be analyzed and assessed.

5.5.3 The civil works construction quality by structures or parts shall be assessed based on the statistical results of self-inspection of construction contractor, random inspection of the supervisor and the third-party inspection, in accordance with the technical requirements for project construction, the contract and the current national and sector standards, taking into account special tests and scientific research results, construction techniques, construction process control, implementation effects, inspection and treatment of anomalies and quality defects, original construction records, project acceptance certificate, and project safety monitoring.

5.5.4 The important concealed works and key locations where quality events occurred and safety hazards exist shall be checked and assessed by random inspection of original project certificates and acceptance documents, construction and supervision work logs, and by the testing results provided by the third-party quality inspection agencies, which are obtained by core drilling, water pressure or grouting test, on-site nondestructive testing, etc.

5.5.5 The new materials and new techniques used in the construction shall be assessed based on the production test results and the implementation effect.

5.5.6 The accuracy and regularity of quality testing results provided by the construction contractors, the supervisor, and the third-party testing agencies shall be analyzed and assessed.

5.5.7 The compliance of the construction quality control standards with the requirements of design and technical standards shall be checked, and the adopted quality control standards not meeting the design requirements shall be assessed based on the review results of the design.

5.5.8 The compliance of factory and site inspection procedures of cement, admixtures, additives, steel and other purchased materials with the quality control procedures and control standard requirements specified for the main raw materials shall be checked; the compliance of quality control links, testing items and frequencies, etc. shall be checked to assess the quality of main raw materials and aggregates by analyzing the testing results from each organization.

5.5.9 The rationality of the concrete mix proportion for construction shall be assessed according to the main concrete mix proportion test results, the development of mix proportion and its adjustment in the implementation, taking into account the project application.

5.5.10 The quality of earth-rock excavation shall be assessed according to the quality control standards of earth rock excavation, and the statistical analysis results of excavation surface testing data such as overexcavation and underexcavation, unevenness, and residual hole rate.

5.5.11 The construction quality of the support works shall be assessed in terms of the quality control standards, control measures, testing items, testing methods, the shotcrete thickness and strength proposed after the statistical analysis, mortar strength and nondestructive or pullout testing results of anchor bolts and anchor piles, as well as the results of anchor cable hole making, anchor cable grouting, anchor pier concrete strength, anchor cable tensioning and grouting, anchor cable dynamometer monitoring, etc.

5.5.12 For concrete dam foundation surface, concrete-faced dam plinth, high toe wall or core wall dam slab, power intake and powerhouse foundation and foundation surfaces of other key parts, the foundation rock mass treatment quality shall be assessed, taking into account the geological defect treatment, exploratory adit and borehole plugging construction and quality acceptance, foundation rock mass sonic testing results, etc.

5.5.13 The quality and effectiveness of foundation consolidation grouting works shall be assessed based on the quality control measures of rock foundation grouting, field grouting tests before construction, selection of grouting materials, grouting methods, construction techniques, control standards, as well as the grouting and foundation treatment in the geologically weak regions with rock breakage and faults, taking into account the occurrence time, location, treatment and observation results of the anomalies during grouting construction, as well as the grouting effectiveness obtained by the sequencing-based statistical analysis, water pressure tests before and after grouting, and sonic testing statistics.

5.5.14 The curtain grouting quality shall be assessed, taking into account the field tests, construction techniques, quality control procedures and anomaly treatment of curtain grouting, and based on the hole drilling and grouting results, inspection hole water pressure test results, geophysical survey and sonic detection results obtained through the statistical analysis by part, row and sequence; the drainage hole construction quality shall be assessed based on the random inspection results of the location, depth, etc. of drainage holes.

5.5.15 The concrete cutoff wall quality shall be assessed based on the testing results of wall shape, concrete compressive strength and elastic modulus, impervious performance, etc. For the foundation concrete cutoff wall on the key parts, the wall quality on the parts where anomalies occurred in the pouring process shall be assessed through the comprehensive analysis of inspection results obtained by coring, water pressure test, sonic detection, and in-hole video.

5.5.16 The concrete construction quality shall be assessed based on the testing and statistical analysis results of fabrication quality and stability, mechanical and overall properties, shape, flow surface, compactness or compaction degree; for the concrete on key parts and all kinds of special concrete such as scour- and wear-resistant concrete, the concrete construction quality shall be assessed based on the analysis of pouring block design, and construction process control conditions.

5.5.17 The concrete temperature control measures and effect shall be assessed based on the testing and statistical analysis results of concrete temperature control measures in low and high temperature seasons, concrete temperature at mixer outlet, placement and pouring temperature by parts and time, concrete water cooling and insulation checks, concrete internal temperature and highest temperature in various periods, etc.

5.5.18 The asphalt concrete construction quality shall be assessed based on the testing and statistical analysis results of the quality of raw materials, mixture preparation quality, as well as placement temperature and thickness, porosity, density, permeability coefficient, etc.

5.5.19 The earth and rock-fill dam filling construction quality shall be assessed based on the testing and statistical analysis results of dam material gradation, filling and roller compaction parameters, compactness, moisture content, etc.

5.5.20 For concealed works such as the joint grouting for dam, contact grouting for bank slope, penstock or steel liner, backfill and consolidation grouting for tunnel liner, backfill, consolidation and joint grouting for plugs of diversion tunnel and bottom outlets, the grouting quality shall be assessed according to the statistical analysis results of grouting and quality check by part or unit; the backfill grouting quality of pressure tunnel shall be assessed based on the results of geophysical prospecting; the quality of joint grouting for dam, contact grouting for bank slopes, and joint grouting for sealed section shall be assessed, taking into account the concrete age and temperature before grouting, burying condition of grouting system, water filling check results before grouting, etc. as needed.

5.5.21 The manufacture quality of penstock and steel lining raw material shall be assessed based on the quality review and spot check results, production welding test results, and statistical analysis results of manufacture dimensional tolerances, shape and location tolerances, weld joint and grouting hole sealing weld and other test records.

5.5.22 The installation quality of penstock and steel lining shall be assessed based on the statistical analysis results of installation dimensional tolerances, shape and location tolerances, weld joint and grouting hole sealing weld and other test records, and the problems identified with installation, corresponding solutions, treatment effectiveness, and their impact on the quality and operation of penstock shall be assessed.

5.5.23 The anti-corrosion construction technique and quality of penstock and steel lining shall be assessed.

5.5.24 The treatment quality and effect shall be assessed based on the analysis of inspection results and treatment effect results of project quality defects. For completion safety appraisal, the treatment effect of problems identified through emptying inspection for the water conveyance system and flood discharge and energy dissipation system shall be assessed.

5.5.25 For the major quality accidents in the construction process, their impacts on the project safety shall be assessed, taking into account the analysis results of accident cause, as well as the investigation and treatment effect, and the countermeasures and suggestions shall be proposed.

5.6 Assessment on Safety Monitoring

5.6.1 The main performance indicators and applicability of important or less-applied new-type instruments and equipment shall be assessed.

5.6.2 The on-site calibration, installation and embedding time and quality, initial value reading, etc. of monitoring instruments and equipment shall be assessed.

5.6.3 The construction quality of monitoring works and the completion and integrity rates of monitoring instruments and equipment shall be assessed; the rationality of construction plan for uncompleted monitoring works shall be assessed in the impoundment safety appraisal based on the impoundment needs.

5.6.4 The effectiveness and timeliness of the analysis and feedback of monitoring results shall be assessed. The retesting accuracy of the deformation monitoring network and its base points stability shall be assessed; for the retesting results of the deformation monitoring network and the calibration results of the last deformation monitoring value before the impoundment, their reliability as the reference values for assessing the operating performance of structures in the impoundment period shall be assessed in the impoundment safety appraisal.

5.6.5 The integrity, continuity and reliability of the monitoring data shall be assessed. If the monitoring of main monitoring items at key parts is interrupted for a long time in the construction period, the impact of missing data on the structure construction quality and its operating performance in the construction period shall be assessed.

5.6.6 The working condition of corresponding structure and slope shall be assessed based on the results of monitoring and analysis of deformation, seepage pressure, stress-strain, temperature, stress on supporting structures, etc.

5.6.7 The inspection system and its performance shall be assessed.

5.7 Assessment on Hydraulic Steel Structures

5.7.1 The assessment on hydraulic steel structures shall include the assessment on the design of trash racks, gates, hoists, shiplifts, etc., and on their manufacture and installation quality, and operating effects.

5.7.2 The assessment on the design of hydraulic steel structures shall meet the following requirements:

1 Assess the layout and type selection of hydraulic steel structures.

2 Assess the material selection, structural layout, water seal arrangement type, support type, slot type, and design parameters of gates and trash racks.

3 Assess the structural layout, parameter selection, and major components type selection of hoists.

4 Assess the strength, stiffness, stability, and safety factor value of the gates and their embedded parts based on the calculation results.

5 Assess the anti-corrosion design for hydraulic steel structures.

6 Assess the major issues of hydraulic steel structures based on the model test results and design review results.

5.7.3 The assessment on the manufacture quality of hydraulic steel structures shall meet the following requirements:

1 Assess the quality of raw materials based on the results of random inspection and re-inspection of steel, welding materials, etc., and assess the performance and quality of outsourced parts based on their results of quality witnessing and quality inspection.

2 Assess the manufacture quality of hydraulic steel structures based on the statistical analysis results of manufacture dimensional tolerances, shape and location tolerances, weld inspection records, and assembly inspection records.

3 Assess the safety and reliability with respect to the problems identified in the manufacture process, corresponding solutions, treatment effectiveness, and their impact on the product quality.

4 Assess the functional indicators of assembled hydraulic steel structures based on test data on factory trial run and tests and joint commissioning of mechanical, electrical and hydraulic parts.

5 Assess the anti-corrosion construction technique and quality of hydraulic steel structures.

6 Assess the pending issues in the factory acceptance and their treatment results.

5.7.4 The assessment on the installation quality of hydraulic steel structures shall meet the following requirements:

1 Assess the damage and deformation of the hydraulic steel structures based on the handover and acceptance records.

2 Assess the installation quality of hydraulic steel structures based on the results of the structural dimensions, shape and location tolerance inspection, weld inspection after installation and corresponding statistical analysis.

3 Assess the installation quality of hydraulic steel structures based on the installation, testing, trial run records and test reports of gates and hoists, such as the fit test of automatic grabbing beam and gate, flushing valve reliability test, static balance, candling inspection, gate and hoist joint commissioning test, and functional test of cleaning grab.

4 Assess the safety and reliability of special equipment such as gate hoists and overhead cranes based on the load test results.

5 Assess the quality, design and manufacture problems identified in the installation process, solutions and treatment effects.

6 Assess the safety and reliability of working power supply and control system of flood release sluice service gates and other flood release facilities, and assess the safety and reliability of working power supply and control system of diversion plugging gates in the impoundment safety appraisal.

5.7.5 The assessment on operating state of hydraulic steel structures shall meet the following requirements:

1 Assess the current state of hydraulic steel structures based on their operating conditions, operating status, and test results. The tests mainly include the load test of the gantry crane and overhead crane, test of service gate operating in flow, test of quick-acting shutoff gate and emergency gate closing in flow, remote control and operation test of gates.

2 Assess the main problems identified in the operation process, solutions and treatment effects, and the technical upgrading results of hydraulic steel structures.

3 Assess the safety and reliability of the long-term operation of the hydraulic steel structures, the reliability of the permanent power supply and backup power supply, and the reliability of the control system.

5.8 Assessment on Turbine-Generator Unit and Auxiliaries

5.8.1 The assessment on turbine (pump-turbine) and its auxiliaries shall meet the following requirements:

1 Assess the type, main design and technical parameters of the turbine, and the structure and materials of its main components.

2 Assess the manufacture and installation quality of turbine draft tube, stay ring, spiral case, runner, head cover, bottom ring, guide bearing, main shaft, servomotor, cylindrical valve and other components based on the manufacture quality inspection and acceptance results, installation quality inspection data and test run results of turbine-generator unit; assess the treatment effect of main problems and defects identified in the manufacture and installation process of turbine components.

3 Assess the energy characteristics, cavitation characteristics and stability of the turbine considering the model test results, the field test and practical operation, and check the completeness of the test items.

4 Assess the main components of the governor system, and the type, main technical parameters, and main component structures and material conformity of the turbine inlet valve.

5 Assess the main components of the governor system, and the installation quality, test, operating conditions of the turbine inlet valve.

6 Assess the main problems identified in the installation and operation process of the governor system and turbine inlet valve, and their treatment effects.

7 Analyze the results of the hydraulic transient design proposed in feasibility study stage, hydraulic transient process contract value, and transient guarantee review results in the bidding and the construction detailed design stage according to the results of load rejection test, assess whether the hydraulic transient design meets the requirements of the project operation, assess whether there are safety hazards in the hydraulic transient process of power station, and provide suggestions.

5.8.2 The assessment on the generator (generator-motor) and its auxiliaries shall meet the following requirements:

1 Assess the type, structure, main technical parameters, and main component materials of the turbine-generator.

2 Assess the installation process and installation quality of the turbine-generator.

3 Assess the installation quality of components such as stator, rotor, bearing, seat and rack.

4 Assess the test results of stator iron loss, stator and rotor voltage withstand, etc.

5 Assess the main problems identified in the installation and test run process and their treatment effects.

6 Assess the design and equipment type selection of excitation system power rectifier bridge, excitation regulator, de-excitation method and excitation method; assess the installation and testing quality of the excitation device and the treatment effect of major defects; assess the performance indexes such as the force excitation factor of the excitation system and the reliability of the excitation system based on the system requirements.

5.8.3 The rationality and completeness of the start-up procedures and test items of turbine-generator unit shall be checked; the results of the tests such as the no-load test, rising current test, rising voltage test, overspeed test, loading test, load rejection test, and continuous operation test of unit with load shall be assessed; the main problems in the test run process and their treatment effects shall be assessed.

5.8.4 The temperature rise, vibration, runout and other indicators of components of turbine-generator unit shall be assessed after it is put into operation.

5.8.5 The rationality of the measures for safe and stable operation based on the stable operation test results of turbine-generator unit shall be assessed; the manufacture and installation quality of the turbine-generator unit shall be assessed according to the installation and test run conditions.

5.9 Assessment on Power Equipment

5.9.1 The grid-connection system design of the power station shall be assessed according to the geographic location and scale. The main electrical connection design shall be assessed according to the requirements for safe, reliable, and flexible operation of the power grid and power station, etc.

5.9.2 The design of the service power supply system shall be assessed. The type of power supply equipment, selection of the key parameters and test results of major installed facilities shall be assessed.

5.9.3 The type, main parameters, layout, quality of manufacture and installation, results of test run of the following generator voltage equipment such as bus, circuit breaker, voltage transformer and arrester cabinet, current transformer, etc. shall be assessed. For pumped storage power stations, the design of the type selection of static variable frequency starting device and phase change switches, their main parameters and arrangement, their manufacture and installation quality, and test run results shall be assessed.

5.9.4 The main transformer type, main technical parameters, layout, quality of manufacture and installation, results of withstand voltage test and partial discharge testing shall be assessed.

5.9.5 The type, main technical parameters, layout, quality of manufacture and installation, testing results of equipment at the high voltage side and switchyard shall be assessed.

5.9.6 The treatment results of the problems identified with the installation and testing of power equipment shall be assessed, the results of various tests shall be checked, and the latest results of preventive tests and the safety and reliability of power equipment in the long run shall be assessed.

5.9.7 According to the operational requirements of the power system, the design of overvoltage protection and insulation coordination for hydropower stations, and the reliability and rationality of the overvoltage protection device configuration and technical parameters selection of lightning arresters shall be assessed.

5.9.8 The design and installation quality of earthing system, and the measured plant earthing resistance, and the contact potential difference, step potential difference, and isolation measures at the important locations shall be assessed.

5.9.9 The design and installation quality of the lighting system shall be assessed.

5.10 Assessment on Control and Protection System

5.10.1 The system reliability shall be assessed according to the operation control method of the computer supervisory control system (CSCS), the construction and operation status of the remote centralized control center and cascade centralized control center, CSCS network structure and configuration, the construction of the network routing and optical fibers and cables for CSCS, and the power supply configuration of control equipment at each level.

5.10.2 The safety redundancy design and facilities for hydropower stations shall be assessed according to the conditions of CSCS realizing normal startup and shutdown, emergency shutdown, and emergency closing of intake emergency gate, taking into account the device independent of CSCS in the central control room to implement emergency shutdown and emergency closing of intake emergency gate, and the configuration of the vibration and swing protection system and the alarm control system of the powerhouse flooding. For pumped storage power stations, the safety design and facilities such as emergency closing of inlet valve and tailrace emergency gate, and locking of inlet valve and tailrace emergency gate shall be assessed.

5.10.3 The reliability of relevant functions of hydropower stations shall be assessed in terms of the success rate of startup and shutdown of hydropower stations and the statistics of the time required for working mode conversion of reversible units, and the implementation of AGC and AVC.

5.10.4 The safety zoning for each system of hydropower stations shall be assessed taking into account the configuration and implementation of safety and protection of the electrical secondary system, safety protection measures for the CSCS and superior scheduling, and other system boundaries, etc.

5.10.5 The design, functions, equipment selection, and installation and testing quality of CSCS, and its main problems identified with operation and results of rectification shall be assessed.

5.10.6 The design, equipment selection, installation and test run quality, main problems identified with operation and their rectification results of the auxiliaries of the generator units and the control system of the plant common equipment shall be assessed.

5.10.7 The design, equipment type selection, installation and testing quality, and the main problems in operation and the rectification results of the gate control system, and its operation monitoring and local and remote control functions shall be assessed.

5.10.8 Whether the configuration and type selection of relay protection equipment comply with the current regulations and specifications, and whether no dead zone can be realized by the relay protection configuration of hydropower stations shall be assessed; the installation and testing quality of relay protection equipment, the main problems identified in operation and their rectification results shall be assessed; the reliability and safety of relay protection equipment shall be assessed taking into account the correct action, false action and action failure of relay protection equipment as well as its intact rate, input rate and correct action rate.

5.10.9 The layout of the DC system, the battery room setting and operating environment and battery endurance time in case of fault shall be assessed according to the configuration of the DC power supply system (including the number of systems, the number of battery packs for each system, and the type of battery), the wiring method of the DC system, the configuration of insulation monitoring equipment, the configuration of battery monitoring equipment, and the results of the DC system capacity reviewed by the statistical analysis on the actual load of the hydropower station during its operation.

5.10.10 The design, equipment type selection, installation and testing quality of the DC system and its main problems in operation and their rectification results shall be assessed according to the requirements of relevant regulations and specifications; the safety and reliability of the DC power supply system shall be assessed taking into account the analysis of battery charge and discharge test results.

5.10.11 The design, equipment type selection, and installation and testing quality of industrial television system and its problems identified with operation and their rectification results shall be assessed; the safety and reliability of the system shall be assessed, taking into account the monitoring range of the front-end monitoring points to the main equipment and safety zone of the hydropower station, as well as the linkage function of the industrial television system, the fire alarm system, alarm control system for powerhouse flooding and other systems.

5.10.12 The setting of synchronizing point, and the safety and reliability of the design and operation of synchronizing method shall be assessed.

5.10.13 The electrical measurement, nonelectrical measurement, and safety and reliability of design and operation of secondary wiring shall be assessed.

5.11 Assessment on Communication System

5.11.1 The channel configuration, reliability, equipment configuration, main technical indicators and functions of system communications shall be assessed.

5.11.2 The reliability, equipment configuration, main technical indicators and functions of cascade communications shall be assessed.

5.11.3 The assessment on the plant communications shall meet the following requirements:

1 Assess the equipment configuration, main technical indicators and functions of plant communications.

2 Assess the test results of self-function and relay functions of program-controlled switches for dispatching and production management.

3 Assess the layout of the integrated communication network.

5.11.4 The assessment on the hydrological telemetry and forecasting system shall meet the following requirements:

1 Assess the design of the communication method, equipment configuration, reliability indicators, power supply system, earthing, main technical indicators and functions of the hydrological telemetry and forecasting system.

2 Assess the test results of main technical indicators of the transceiver such as the power, bit error rate, transmission loss, system's normal operation rate, and the time of data acquisition, transmission and processing.

5.11.5 The assessment on different communication methods shall meet the following requirements:

1 Assess the test results of main technical indicators such as the transceiver power, system loss, shaking, bit error rate, and splice loss when the optical fiber is used for the system communications, cascade communications, and plant communications.

2 Assess the test results of the power, receiver sensitivity, thermal noise bandwidth, transhybrid loss, and main technical indicators of the transceiver used for the power line carrier terminal, the barrage width of wave traps, the coupling capacitors, and the high-frequency cables and working frequency bands of combined filters, when the power line carrier communications are used for system communications.

3 Assess the test results of main technical indicators such as receiver power, frequency band, antenna feed system, antenna elevation angle and in-band interference when the satellite is used for the system communications, cascade communications, and plant communications.

5.11.6 The assessment on the communication power supply shall meet the following requirements:

1 Assess the reliability, main technical indicators and functions of power supply system.

2 Assess the main technical indicators such as the capacity and charge and discharge of the power supply system.

5.11.7 The lightning protection and earthing equipment of communication equipment, power supply system, satellite ground station and telemetry station of the hydrological telemetry and forecasting system, relay station and central station shall be assessed.

5.11.8 The main problems identified in the installation, testing and operation of system communications, cascade communications, plant communications, hydrological telemetry and forecasting system and power supply system, and their rectification results shall be assessed.

5.12 Assessment on the Rest Ancillary Equipment and Systems

5.12.1 The conformity of the model, specification, main technical parameters and layout of the lifting equipment in the powerhouse shall be assessed; the manufacture and installation quality of lifting equipment in the powerhouse shall be assessed according to the equipment manufacturing quality inspection and acceptance results, installation quality inspection data, equipment test data and on-site operation, in comparison with the manufacturer's technical documents; according to the load test results of the lifting equipment in the plant, the use license conditions of special equipment and the on-site use conditions, the safe operating conditions of the lifting equipment, as well as the rectification results of the main problems and defects identified in the installation and operation of the lifting equipment shall be assessed.

5.12.2 The model and specification, main technical parameters and layout of main equipment shall be assessed for various systems such as the cooling water supply system, medium- and low-pressure compressed air system, turbine oil system, insulate oil system, hydraulic power monitoring and measuring system, dewatering system, drainage system for powerhouse and dam; the installation quality of each system shall be assessed according to the installation quality control and test results; the effect and reliability of each system shall be assessed, taking into account the actual operation, and the rectification results of the main problems and defects found in the installation and operation of each system shall be assessed. The heating ventilation and air conditioning (HVAC) system shall be assessed taking into account the test.

Appendix A Contents of Safety Appraisal Program

A.1 Project Overview

A.2 Scope and Tasks of Safety Appraisal

A.3 Main Basis for Safety Appraisal

A.4 Main Content of Safety Appraisal

A.5 Safety Appraisal Method

A.6 Safety Appraisal Plan

A.7 List of Required Documents and Information Supplied by All Parties and Relevant Requirements

A.8 Attachment: Preparation Requirements for Self-inspection Report of All Parties for ××× Project (Special, Impoundment or Completion) Safety Appraisal

Appendix B Contents of Safety Appraisal Report

B.1 The Safety Appraisal Report May Be Cataloged According to Disciplines, Structures (Dam, Powerhouse, Flood Discharge Tunnel, etc.), or Their Combinations

B.2 The Reference Contents Arranged by Discipline Are as Follows

B.2.1 Overview of Safety Appraisal

B.2.2 Overview of Hydropower Complex

B.2.3 Assessment on Project (or Project Impoundment and) Flood Control and Operation Dispatching

B.2.4 Assessment on Engineering Geological Conditions

B.2.5 Assessment on Hydraulic Structure Design

B.2.6 Assessment on Civil Works Construction Quality and Penstock Manufacture and Installation Quality

B.2.7 Assessment on Safety Monitoring

B.2.8 Assessment on Hydraulic Steel Structures

B.2.9 Assessment on Electromechanical Engineering

B.2.10 Major Problems and Assessment on Their Handling Results

B.2.11 Conclusions and Recommendations

B.2.12 Attachments and Attached Drawings

1 Attachment 1: Table of Main Characteristics of the Project

2 Attachment 2: Signature Form of Expert Team Members for ××× Project (Special, Impoundment, or Completion) Safety Appraisal

3 Attachment 3: Documents and Information for ××× Project (special, impoundment, or completion) Safety Appraisal

4 Attached Drawings

Appendix C Required Documents and Information Supplied by All Parties

Table C Required documents and information supplied by all parties

Documents	Special safety appraisal	Impoundment safety appraisal	Project completion safety appraisal
Ⅰ. Submittals			
1. Self-inspection report of project construction management of project owner (including previous quality supervision reports and on-site inspection results)	√	√	√
2. Third-party inspection report	√	√	√
3. Design self inspection report (including attached drawings)	√	√	√
4. Supervision self-inspection report (including manufacture supervision self-inspection report)	√	√	√
5. Construction self-inspection report (including manufacture self-inspection report)	√	√	√
6. Operation self-inspection report	○		√
7. Processing and analysis report of safety monitoring data	√	√	√
8. Other reports required by the safety appraisal expert team	√	√	√
Ⅱ. Documents for reference			
1. Approval and filing documents related to project construction	√	√	√
2. Feasibility study reports and design change reports and review comments	√	√	√

Table C (*continued*)

Documents	Special safety appraisal	Impoundment safety appraisal	Project completion safety appraisal
3. Model test and other special research reports and relevant review comments or advisory opinions	√	√	√
4. Contract documents, construction drawings, design notices, etc.	√	√	√
5. Stage acceptance reports and appraisals for project and acceptance reports and appraisals for works	√	√	√
6. Documents for acceptance certification	√	√	√
7. Original construction records	√	√	√
8. Monthly reports, annual reports and interim reports of supervision and safety monitoring	√	√	√
9. Consultation report on major issues	○	○	○

NOTE "√" denotes mandatory, and " ○ " denotes optional as needed.

Appendix D Preparation Requirements for Self-Inspection Report

D.1 Layout

D.1.1 Front Cover (See Figure D.1.1)

D.1.2 Inner Cover (See Figure D.1.2)

D.1.3 Contents

D.1.4 Main Body

D.1.5 Attachments

D.2 Font Size and Font Type

D.2.1 Headings in the Main Body

1 Chapter and section headings are in 16-point SimHei and KaiTi font respectively

2 The item headings are in 14-point SimHei font

D.2.2 Content in the Main Body

1 The text is in 12-point SimSun font.

2 The words in tables may be in either 10.5-point or 7.5-point SimSun font.

D.3 Paper and Layout

1 A4 white offset paper (70 g or above).

2 The left margin is 2.65 cm, the right margin is 2.65 cm, and the top and bottom margins are 2.54 cm.

3 The chapter and section headings are centered, and two blank spaces are indented for item headings.

D.4 Printing

Printed on both sides except for the attached drawings, photocopies, etc.

D.5 Cover Page

The front cover of the self-inspection report is stamped with the official seal of the preparation organization, and each input item shall be signed by the responsible person.

××× Project

(Special, Impoundment or Completion)

Safety Appraisal Self-Inspection Report (No.: ××)

(15-point FangSong GB 2312)

××× Project

(Special, Impoundment or Completion)

Safety Appraisal

(18-point FangSong GB 2312)

(Development Management, Design, Supervision, Construction and Installation, or Operation)

Self-inspection Report

(36-point FangSong GB 2312)

(Subtitle)

(16-point FangSong GB 2312)

Development organization (official seal)

Date: ××××

(15-point FangSong GB 2312)

Figure D.1.1 Self-inspection report front cover

××× Project
(Special, Impoundment or Completion)

Safety Appraisal

(18-point FangSong GB 2312)

(Development Management, Design, Supervision, Construction and Installation, or Operation)

Self-inspection Report

(Subtitle)

(24-point FangSong GB 2312)

Approved by:

Checked by:

Reviewed by:

Prepared by:

(16-point FangSong GB 2312)

Figure D.1.2 Self-inspection report inner cover

Explanation of Wording in This Specification

1 Words used for different degrees of strictness are explained as follows in order to mark the differences in executing the requirements in this specification.

1) Words denoting a very strict or mandatory requirement:

"Must" is used for affirmation; "must not" for negation.

2) Words denoting a strict requirement under normal conditions:

"Shall" is used for affirmation; "shall not" for negation.

3) Words denoting a permission of a slight choice or an indication of the most suitable choice when conditions permit:

"Should" is used for affirmation; "should not" for negation.

4) "May" is used to express the option available, sometimes with the conditional permit.

2 "Shall meet the requirements of..." or "shall comply with..." is used in this specification to indicate that it is necessary to comply with the requirements stipulated in other relative standards and codes.